DATE DUE

Wind and Air Pressure

Alan Rodgers and Angella Streluk

Heinemann Library
Chicago, Illinois

© 2003 Reed Educational & Professional Publishing
Published by Heinemann Library,
an imprint of Reed Educational & Professional Publishing,
Chicago, Illinois
Customer Service 888-454-2279
Visit our website at www.heinemannlibrary.com

Designed by Storeybooks
Originated by Ambassador Litho Limited
Printed in Hong Kong/China

07 06 05 04 03
10 9 8 7 6 5 4 3 2 1

Library of Congress Cataloging-in-Publication Data
Library of Congress Cataloging-in-Publication Data

Rodgers, Alan, 1958-
  Wind and air pressure / Alan Rodgers and Angella Streluk.
      v. cm. -- (Measuring the weather)
Includes bibliographical references and index.
Contents: What is wind? -- Wind direction -- Land and sea -- Wind chill -- Wind and weather -- Measuring wind strength -- The beaufort wind scale -- Using technology to monitor the wind -- Extreme winds -- Global winds -- What is air pressure? -- Barometers -- Highs and lows.
  ISBN 1-58810-690-X (HC) -- ISBN 1-40340-130-6 (PB)
  1. Winds--Juvenile literature. 2. Winds--Speed--Measurement--Juvenile literature. 3. Atmospheric pressure--Juvenile literature. [1. Winds. 2. Atmospheric pressure.] I. Streluk, Angella, 1961- II. Title.
  QC931.4 .R64 2002
  551.51′8--dc21
                          2002004015

Acknowledgements
The Publishers would like to thank the following for permission to reproduce photographs:
Stone, pp. 4, 10, 21; Trevor Clifford Photography; pp. 6, 15, 19, 26; Photocopy, p. 7; Robert Harding, p. 8; FLPA, p. 9; Bruce Coleman Collection, p. 14; The Met Office, p. 16; D. Offiler, p. 18; Science Photo Library, p. 24; Corbis, p. 25; Alan Rodgers, p. 27.

Cover photographs reproduced with permission of Photodisc and Tudor Photography.

Our thanks to Jacquie Syvret of the Met Office for her assistance during the preparation of this book, and to Amington Heath Community School.

You can find words in bold, **like this,** in the Glossary.

Contents

What Is Wind?

Wind is one of the most important features of our weather. Wind can bring rain, and it can cause hot or cold weather. It is one of the main features of storms. In this book, you will find out how the wind works and what effect it has on the weather.

When we feel the wind on our faces, we are feeling the movement of air. Wind is an **air mass** that moves from one area to another. Air from high **air pressure** areas always flows to low air pressure areas. The greater the difference in pressure between the areas, the stronger the wind will be.

Why is the pressure of air different in different areas? Changes in air pressure are linked to the amount of sun an area gets. If an area gets lots of sun, the air becomes hot, rises, and leaves a space that needs to be filled. Cooler air comes in to fill this space, causing a wind. The Sun is therefore the driving power behind air pressure and the wind.

Wind plays a part in some enjoyable activities. Many children enjoy flying kites. Adults also enjoy sports that depend on the wind. These include hang-gliding and sailing, in which the direction of the wind is very important.

Weather and climate

Many people need to know what the weather will be. Sometimes wind can be very dangerous. It can cause damage and injury to property and people. So scientists called **meteorologists** measure the weather and try to predict what it will do in the future. Meteorologists around the world try to record the information in the same way so everyone can share and understand it. They show some of their weather information in weather reports, using sets of symbols like those shown in the diagram below. The symbols used are usually ones that can easily be understood by anybody. Sometimes they are not quite the same, even in the same country, but if you know what one set means, it is usually easy to see what another set means too.

In any one place, the weather changes a lot over a long period of time. By recording these patterns in the change of weather, weather watchers can describe the **climate** of an area. Climate is a word used to describe the patterns of the weather in a place over a large number of years.

Be careful!

Do not look directly at the Sun when studying the weather. Also, never take shelter under trees during a thunderstorm, as they may be hit by lightning.

These are the symbols used by meteorologists throughout the world to represent wind speeds. They are added to **station circles** (like the one in the diagram) to show the wind direction in a particular place.

Knots are a unit of measurement used to describe wind speed; 1 knot equals 1.15 miles per hour.

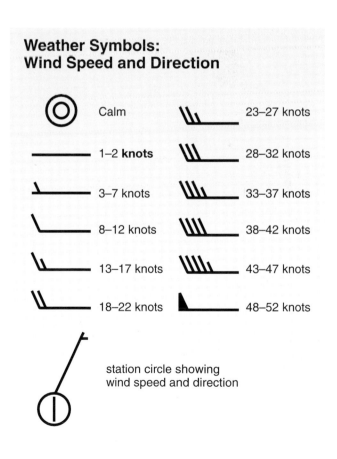

**Weather Symbols:
Wind Speed and Direction**

Calm		23–27 knots
1–2 **knots**		28–32 knots
3–7 knots		33–37 knots
8–12 knots		38–42 knots
13–17 knots		43–47 knots
18–22 knots		48–52 knots

station circle showing
wind speed and direction

Wind Direction

Knowing the direction that the wind comes from is an important part of measuring the weather. We can watch flags and trees to see which way the wind is blowing. However, scientists need to measure the wind in a more accurate way. One of the simplest and oldest weather recording devices is the **weather vane,** like the one shown below.

The pointed end of the weather vane is the most important part. It tells us which way the wind is coming from—not where it is going to. The wind blows the sail end (which is larger than the pointer end) until the vane is in line with the wind and does not block the wind anymore. Sometimes the wind can change direction and catch the sail from a different direction. If the wind changes in a clockwise direction, it **veers.** If it changes in an counterclockwise direction, it **backs.** If the wind keeps changing direction like this, watch the vane for 30 seconds. Then decide which way the vane has pointed for most of that time.

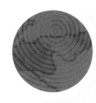

Try this yourself!

Ask an adult to help you make a simple weather vane.

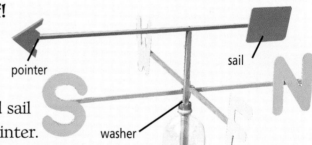

- Make a cardboard sail fixed to a stick pointer.
- Mount the pointer on an upright piece of wood. Use a washer so that it can turn freely. Be sure to balance the pointer properly, as the sail end will be heavier than the pointer end.
- Make labels for north, south, east, and west.
- Weight a plastic bottle with water or stones to make a base.
- Place the vane carefully so that nothing obstructs a free flow of wind around it. Set it up in an open space away from obstacles such as hedges, fences, and buildings.
- Use a compass to turn the direction labels so that they point correctly.

Electronic devices

Wind direction can also be measured by electronic devices. Some are labeled with the directions of the compass and show the direction with a pointer. Other devices use a digital display that shows the wind direction as degrees of a circle. For example, north is 360°, east is 90°, south is 180°, and west is 270°.

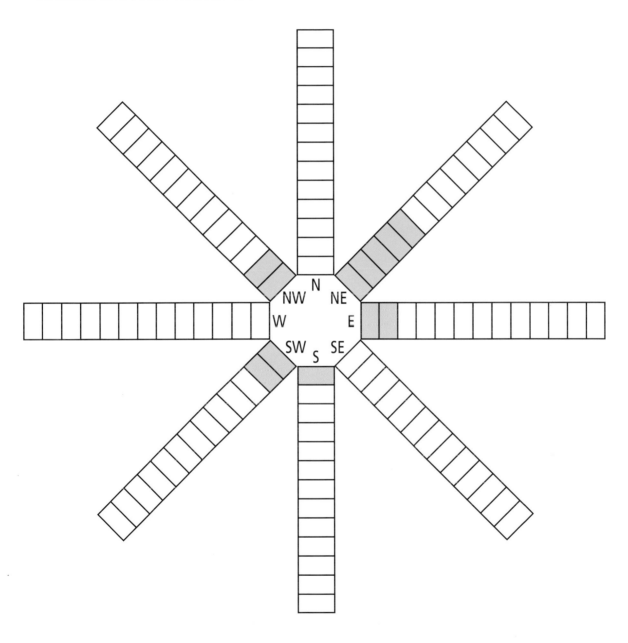

Use the data from your weather vane to fill in a "wind rose." Color in a square each day along the line for the wind direction of that day. After several weeks of observation you will see if there is a **prevailing** wind. Here, the prevailing wind is from the northeast. Remember, when there is no wind, no direction can be recorded.

Land and Sea

People often go to the beach on hot sunny days to enjoy the cool ocean breezes. These breezes are caused by the difference in the temperature of the land and the sea. The ground and air heat up at different speeds. The ground and sea also heat up at different speeds. This brings about winds called sea breezes and land breezes.

Sea breezes

Sea breezes are light winds that travel from the sea to the land. During the day, land warms up quickly in the Sun. The sea warms up more slowly. In the summer, the difference between the temperature of the land and sea can be large. This causes the air to move in a large **cycle**.

1. During the day, the warm land heats the air above it, which then rises. This causes an area of low air pressure, which is ready to be filled by more air.
2. The cooler air from the sea comes in to take the place of the warm air that has risen. This is the sea breeze we feel.
3. The warm air, high in the sky, goes out to sea.
4. At sea, the warm air is cooled and sinks back down. The cycle continues.

The biggest difference in temperature is during the afternoon, and this is when sea breezes will be strongest.

Land breezes

Land breezes are the opposite of sea breezes. At night, the land cools down quite quickly compared to the sea. The air above the sea is warmer than that over the land, and it rises, creating an area of low **air pressure.** This causes a breeze from the land to the sea.

The difference between land and sea temperatures at night is not as great as during the day. This means that land breezes are not as strong as sea breezes, but ships near the coast may notice them. These breezes also occur at the edge of large lakes. These are known as lake breezes.

1. At night, the sea cools down quite slowly compared to the land. The air above the sea is warmed by the sea. It rises, creating an area of low air pressure.
2. Cool air from the land moves in to fill the space left by rising air over the sea.
3. The warm air from high above the sea moves towards the land.
4. The warm air sinks as it cools over the land. It takes the place of the air that went out to sea.

This cycle continues as long as there is a difference in temperature between the sea and land.

Wind Chill

Sometimes a day might feel really cold, even though the temperature is not that low. We especially notice that our face gets cold when the wind blows on it. Heat given off by the skin warms the air close to our face, but when the wind blows hard, it drives away the warmth. The way the wind cools things down like this is known as wind chill.

Two places close to each other, but at very different heights, will have very different temperatures. The people standing at the foot of the mountain do not need coats. At the top of the mountain, at the same time, the hiker above needs to wear more clothing. The extra clothes protect him against the combined effects of wind chill and being at a higher **altitude**.

High and low

On a summer day it might feel quite warm at the bottom of a mountain. People may be wearing light clothing. If they travel up the mountain, however, they will find it feels much colder. This is partly because temperature decreases as height decreases, and party because of wind chill. The strong winds found on tops of hills and mountains blow heat away from people and make it feel colder than the temperature actually is. People have died from the cold because they thought a warm day at the bottom of a mountain meant that it would feel the same at the top.

Figuring out wind chill

In cold **climates,** people who spend a lot of time outside use a Wind Chill Equivalent Temperature Chart to find out how cold it is going to feel. The difference between the real temperature and what the temperature feels like is known as the wind chill factor. For example, an air temperature of 30° F (-1.1°C) with a wind speed of 10 mph (16.1 kph) will mean that the wind chill equivalent temperature will be 21° F (-6.1° C). As a general rule, the stronger the wind, the colder it will feel.

		Air Temperature							
		30° F	20° F	10° F	0° F	-10° F	-20° F	-30° F	-40° F
Wind Speed (mph)	5	25	13	1	-11	-22	-34	-46	-57
	10	21	9	-4	-16	-28	-41	-53	-66
	15	19	6	-7	-19	-32	-45	-58	-71
	20	17	4	-9	-22	-35	-48	-61	-74
	25	16	3	-11	-24	-37	-51	-64	-78
	30	15	1	-12	-26	-39	-53	-67	-80
	40	13	-1	-15	-29	-43	-57	-71	-84

Wind chill can be figured out using this table. To read it you need to know the temperature (be sure to take the reading in a place out of the wind) and the strength of the wind. Trace down from the correct temperature and across from the wind speed. The number where they meet is the temperature that you will feel.

Wind and Weather

When forecasting the weather, you will need to know where you are in relation to the land and sea around you. Whether you are near the ocean or in the center of a large area of land will affect the weather you get. Generally, if a wind is **maritime** and has blown over the sea, it will collect a lot of moisture from the sea. Air that is warm will hold more moisture than cooler air. If a wind has come from a warm place, then it will bring warm weather. If a wind is **continental** and has blown over land, then the wind will generally be drier. Wind that has blown from a cold part of the world, such as one of the polar regions, will be colder.

The four main types of air masses		
	Warm	**Cold**
Land	Continental tropical (cT)	Continental polar (cP)
Sea	Maritime tropical (mT)	Maritime polar (mP)

The chart above shows the professional terms for **air masses.** These terms depend on where the wind comes from. Which winds are likely to bring rain? When might dry weather be brought by wind? The map gives an example showing the expected types of wind for four wind directions affecting London in the United Kingdom. Can you tell which winds should have which terms from the chart?

Mountain and valley breezes

In many places the shape of the land causes different winds. Mountain and valley breezes are examples of this. During the day, sunlight warms the sides of a valley. This causes air to rise up the slopes. This makes a gentle rising breeze, known as a valley breeze. At night, the mountain slopes cool quickly, and they cool the air near them. This cool air is dense, so it sinks down into the valley. This movement of air is called a mountain breeze.

Sometimes air is warmed when it flows down the sides of mountains. One example of this is a wind called the Chinook, which comes down the eastern slopes of the Rocky Mountains in North America. Because this wind is dry and warm, it can melt snow. In New Zealand, the Canterbury Northwester is a similar type of wind. It blows down the mountains, taking warm air to the Pacific Ocean.

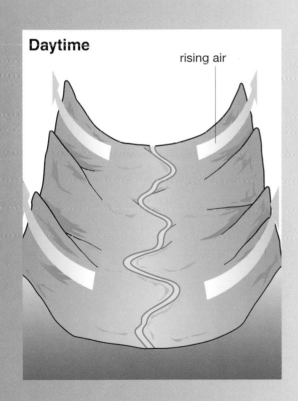
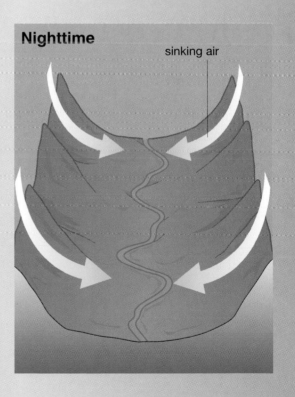

Daytime

rising air

Nighttime

sinking air

As the air warms up during the day, it rises from the valleys and up onto the mountains. This creates a valley breeze. As the air cools at night, it sinks down the hills and into the valleys. This creates a mountain breeze.

Measuring Wind Strength

Knowing the correct wind speed can be very important. For example, people parachuting for the first time are only allowed to jump if there is no more than a gentle wind. Airplane pilots also need to know how strong winds are when they are taking off and landing, so that they can adjust their controls.

Wind measuring devices are called **anemometers**. There are many different types of anemometers. Some will also tell you the direction of the wind if they have wind vanes attached to them. Other anemometers may also show the wind speed.

How to measure wind speed

To measure wind speed, stand in an open space and make sure that there are no buildings or trees nearby. Stand with the strongest wind on your left and hold out your wind measuring device with your arm out straight.

Many modern boats have an automatic **weather station** that includes an anemometer. This is especially important if the boats are going on long journeys.

This will help to keep your body from disturbing the wind. Watch the indicator for a full minute and make a note of the readings every fifteen seconds. Find the average by adding up your four readings and then dividing the sum by four. When the wind blows harder for a short time it is called a gust. Gusts should also be recorded. They will give readings that are higher than most of the other readings taken at the same time.

It is important that an anemometer is used away from any obstructions, such as trees or buildings. Obstructions may speed up or slow down the wind before it gets to the anemometer, giving incorrect readings.

Did you know?

Wind speed can by recorded as:
- feet or meters per second
- miles or kilometers per hour
- **knots** (nautical miles per hour; a nautical mile is 6,080 feet)

Once the speed is known, the correct **Beaufort Scale** reading (see pages 16–17) can be figured out.

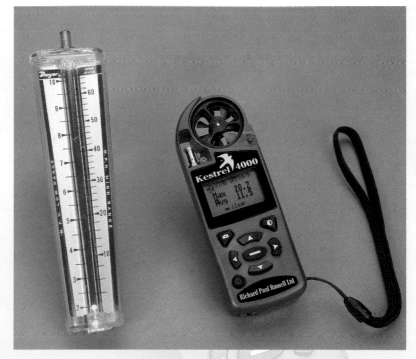

A handheld anemometer can be very accurate. The wind meter shown in this picture (at left) is a small and light anemometer. Modern electronic portable weather monitors like the one on the far right can record wind gusts and average wind speed. They can also calculate wind chill.

The Beaufort Wind Scale

Up until about 150 years ago, most ships relied on the wind to make them move. Sailors needed to know what the wind was like so they could decide which sails to put up. If the ship had too many sails up, it would blow over. If it had too few sails up, the ship would not be able to reach its fastest speed. Unfortunately, there were no standard descriptions of wind strengths at that time. Then, in 1805, Sir Francis Beaufort made a scale to describe what the sea looked like when the wind was at different strengths. He did this mainly by describing how high the waves were. A similar scale was later made for use on land. This is what many weather watchers use today.

Reading the Beaufort scale

The **Beaufort Scale** tries to make it easy to describe wind strength accurately. Each strength has a number, a description, and the wind speed. There are also a few words to describe what the viewer can see, which help in choosing the correct number on the scale. The wind speed is given in miles per hour (mph) and kilometers per hour (kph) for weather forecasts, but for ships and aircraft it is given in **knots.** One knot is equal to 1.15 mph (1.85 kph).

Sir Francis Beaufort (1774–1857) was born in Ireland. He was the first person to use descriptions of the sea to figure out the strength of the wind.

Beaufort Scale		
Number and Description	Features	Air Speed mph (kph)
0 Calm	Smoke rises vertically; water smooth	less than 1 (less than 1)
1 Light air	Smoke shows wind direction; water ruffled	1–3 (1–5)
2 Light breeze	Leaves rustle; wind felt on face	4–7 (6–11)
3 Gentle breeze	Loose paper blows around	8–12 (12–19)
4 Moderate breeze	Branches sway	13–18 (20–29)
5 Fresh breeze	Small trees sway; leaves blown off	19–24 (30–39)
6 Strong breeze	Whistling in telephone wires	25–31 (40–50)
7 Near gale	Large trees sway	32–38 (51–61)
8 Gale	Twigs break from trees	39–46 (62–74)
9 Strong gale	Branches break from trees	47–54 (75–87)
10 Storm	Trees uprooted; weak buildings collapse	55–63 (88–102)
11 Violent storm	Widespread damage	64–72 (103–116)
12 Hurricane	Widespread structural damage	above 72 (above 116)

The Beaufort scale means that you don't need an **anemometer** to estimate the wind strength. You can observe your surroundings. If the leaves on the trees are rustling, then you may note that it is a level 2 on the scale: a light breeze of about 5 mph (8 kph) on average. If you can see large branches moving on the trees then it may be a number 6 on the scale. This means the wind is blowing at 25 to 31 mph (40 to 50 kph).

Using Technology to Monitor the Wind

It is possible to use simple electronic equipment to measure the wind. **Raw data** is most useful when it is analyzed to make charts and graphs. The simple data you collect can then help show you **weather trends** or how weather changes over time.

You will often see **anemometers** at airports, by railroads, on high bridges, near ports, and on tall buildings. Many are linked to electronic monitoring systems and computers. These systems can collect the data, save it, and display the wind speed. Some anemometers are placed out at sea and they may send their data, by satellite, to **monitoring stations.**

Using an anemometer

You can use an anemometer and a computer to find the best place to put wind monitoring equipment. Different wind speeds can be recorded in different locations, even within small areas. The effects of buildings, trees, and other obstacles can affect the force of the wind.

Although it is easier to use anemometers on land, the more useful ones may be those out at sea on ships. These are very important because they monitor incoming weather.

These obstructions do not always slow the wind down. When there are small gaps between buildings, wind speed can actually increase because the gaps squeeze the wind through faster. People who live in cities with tall buildings know about this. Weather watchers normally position their anemometers 33 feet (10 meters) above the ground to avoid the effects that buildings and other obstacles have on the wind.

These children are investigating the difference in wind strength in different locations. They could change the height of the anemometer several times in one location to see if the wind speed is different.

Try this yourself!

Investigate wind speed by measuring the wind strength in different locations.

- Select a few places where you think the wind speed will be different. Mark these on a map.
- Measure the wind speed for a set period of time in the different locations.
- Enter the data into a graphing program.
- Look at the graphs to see what similarities and differences you can spot.

If you don't have a graphing program, ask a teacher to show you how to draw graphs of your data.

Extreme Winds

Terrible storms happen all over the world. They are called different names in different places. Hurricanes are storms starting in the warm southern part of the North Atlantic Ocean or eastern North Pacific Ocean. Similar storms in the China Seas and western North Pacific area are called typhoons. In India and Australia, these storms are called **cyclones.** All these storms are known internationally as tropical cyclones. Tropical cyclones happen in the warmer seasons, when the conditions are right. Wind and rain, combined with a drop in **atmospheric pressure,** create a revolving storm which can cause terrible damage. Cyclones are 300 to 500 miles (500 to 800 kilometers) across, with a minimum wind speed of 64 **knots.**

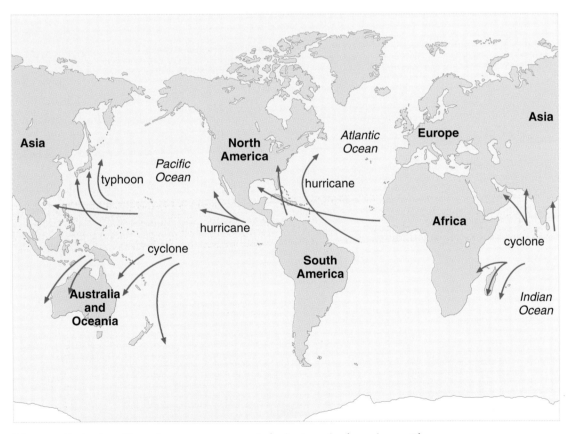

The most destructive storms occur mainly in tropical regions. They usually start over the oceans. Data from satellites is used to give warnings, wherever possible, of approaching storms. The blue arrows on this map show where the various types of storms generally occur.

Tornadoes

Tornadoes are caused by moist, warm air rising rapidly at about 100 miles per hour (160 kph). They are produced by strong thunderstorms. The movement of air between different areas of pressure causes winds at the top of the storm to spin. They develop into a twisting column of wind that picks up dust and debris from the ground. The dust and debris travel up and down the column and make it look dark.

There is a big difference in **air pressure** inside the column compared to outside it. This difference means that if it passes directly over a building, the building can be damaged. Along with the very strong winds, this air pressure difference makes a tornado very destructive. Tornadoes are so strong that they are measured on a special scale called the Fujita Scale for Damaging Wind. Hurricanes are measured on the Saffir-Simpson Hurricane Damage Potential Scale.

Waterspouts and dust devils are the smaller sea and land versions of these storms. Waterspouts do not rotate as quickly as tornadoes. They happen only over the sea. Dust devils form on clear, hot days when warm air rises from a hot surface. They are usually small and last only for a brief time. In Australia these are known as "willy-willies."

This whirling spiral of wind is very dark because it is full of debris that it has sucked up. Nobody would want to photograph a tornado from very close, as it is extremely dangerous.

Global Winds

So far we have looked at winds that happen close to land. But high in the **atmosphere** there are other winds. The main cause of these high **altitude** winds is the difference in temperature in different parts of the world. For example, hot air rises at the **equator.** Cooler air takes its place, and this creates wind.

This difference in temperature creates a regular system of air movements called the Hadley cells. For example: in the tropics, winds always flow westward, creating what is known as the trade winds. For hundreds of years sailors have known about these winds and used them to plan their voyages. Some areas, especially around the equator, never have these winds and are known as the doldrums.

Jet streams

These "rivers of wind" flow in particular directions and areas of the globe and their speeds vary. They were discovered during World War II when airplanes began to fly at higher altitudes. The jet streams flow in a generally easterly direction, although they travel in wavy patterns.

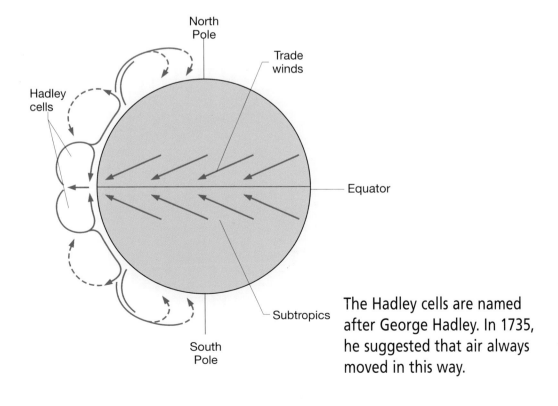

The Hadley cells are named after George Hadley. In 1735, he suggested that air always moved in this way.

Meteorologists use **weather balloons** to measure these winds. This helps predict the jet stream's effects on the weather. It also provides information for airplanes that fly high up in the sky.

El Niño

One of the trade winds makes water move in the South Pacific Ocean. A current of surface water is carried west. This allows cold water to rise up from deeper down in the ocean. The cold water brings up nutrients that feed ocean life off the west coast of South America. Every few years the trade winds weaken, changing the temperature and rainfall patterns over a large area. This also means that the direction of the South Pacific equatorial current changes. This prevents the rising of the cold water and nutrients, and it also brings a current of warm water to the coast of Peru. When this happens, the **climate** in a large area is affected. This effect is known generally as El Niño.

El Niño

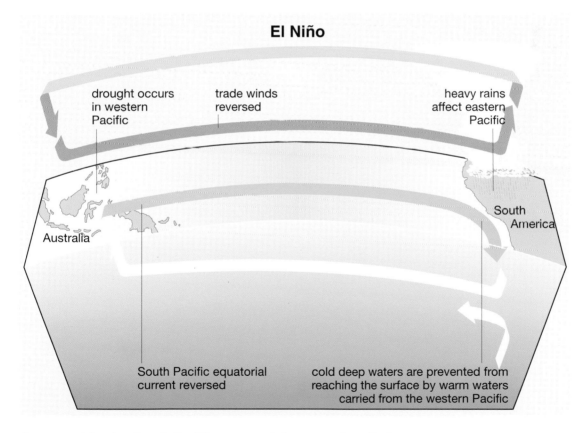

A reversal in the South Pacific equatorial current is called an El Niño. Because weather events are usually connected, the reversal has an effect on a wider area. This diagram shows what happens during an El Niño.

What Is Air Pressure?

Air pressure is the weight of air pressing down. Did you know that 10.8 square feet (1 square meter) of air weighs 22,046 pounds (10,000 kilograms) on the ground? Imagine this much weight on your head! You don't feel this weight because you have air inside you that is at the same pressure as the air outside.

You can feel this air inside you if you travel in an airplane. The air pressure changes quickly, and the air pressure on each side of your eardrum does not have time to readjust. Eventually your ear will "pop" as the pressure suddenly becomes the same.

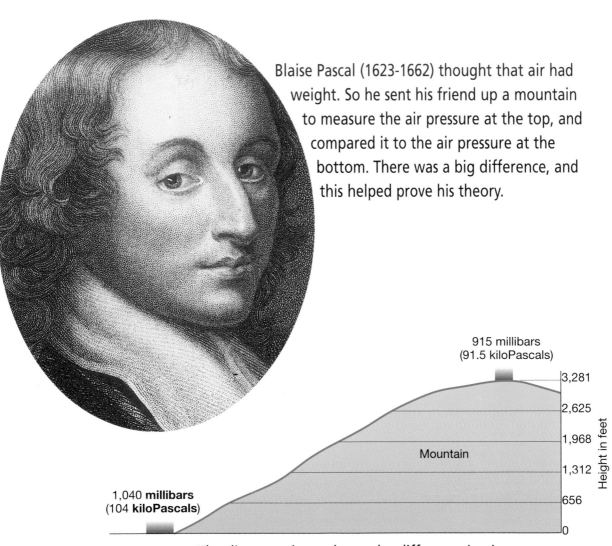

Blaise Pascal (1623-1662) thought that air had weight. So he sent his friend up a mountain to measure the air pressure at the top, and compared it to the air pressure at the bottom. There was a big difference, and this helped prove his theory.

915 millibars
(91.5 kiloPascals)

1,040 **millibars**
(104 **kiloPascals**)

Mountain

Height in feet

3,281
2,625
1,968
1,312
656
0

The diagram above shows the difference in air pressure at the top and bottom of a mountain that is 3,281 feet (1,000 meters) high. There is much less pressure at the top.

Try this yourself!

- Hang a coat hanger up and make sure it balances evenly.
- Find two identical balloons.
- Blow up one of the balloons.
- Hang a balloon on each end of the coat hanger.
- The end with the blown-up balloon goes down. This shows that air has mass (weight).

Different pressures

The air in different parts of the world has different pressures. Weather forecasters show the difference in air pressure on their maps by drawing lines that connect areas with the same air pressure together. These lines are called isobars. In simple terms, most low-pressure areas are caused by rising warm air. Most high-pressure areas are caused by falling cold air. Air from high-pressure areas moves to take the place of the rising air in low-pressure areas. We feel this air movement as wind.

Airplanes use air pressure to tell how high they are. Pilots need to regularly contact local air traffic controllers to find out the current air pressure of an area. They then need to adjust their instruments to the correct reading. They can then fly accurately, avoiding obstacles such as mountains.

Barometers

You can measure **air pressure** using a barometer. **Mercury** barometers measure the effect of air pressure on a tube of mercury. This is read using a scale. Another type, which is more convenient and popular, is called an **aneroid** barometer. This works by measuring the pressure of the air on a sealed unit which has had much of its air sucked out.

Aneroid barometers can be placed indoors so they can be read conveniently. Although they are indoors, they measure the pressure of the **atmosphere** outside. However, they should not be placed near any strong drafts. The drafts will change the pressure and cause the barometer to give incorrect readings.

Many barometers have descriptions of the weather written by the scales. These are not as useful as they seem. They only provide a very rough guide to future weather, or tell you about the weather as it currently is. When it is stormy, you do not need a barometer to tell you!

A barometer looks like a clock but it measures air pressure, not time. Most scientific barometers have their scale in millibars (mb for short) or kiloPascals (kPa), but some have it in inches.

Using a barometer

Read the barometer by looking at the pointer and making a note of its position. Gently tap the face of the barometer and see how much it moves, and whether it is rising or falling. The average reading should be about 1,013 **millibars** or 101.3 **kiloPascals.** A very low reading might be 960 millibars or 96 kiloPascals. A very high reading might be 1,040 millibars or 104 kiloPascals. When you have made a few readings, every three hours or so, you can make some predictions about the approaching weather, using the table below.

Barometer Readings and the Weather They Predict	
Pointer movement	Weather prediction
Rapid fall in air pressure	Stormy weather
Steady—little change in air pressure	Weather the same as before
Steady rise in air pressure	Nice weather

This type of barometer is called a barograph. A pen on the end of an arm records the air pressure on a revolving chart. This chart will record a week's data. The line shows the rising and falling air pressure.

Highs and Lows

You can find out what will happen to the weather from weather forecasts on the radio, television, in newspapers and magazines, and on the Internet. Much of the information is shown on maps called charts. The charts that show **air pressure** are called isobar charts.

You will often see the lines called isobars on a weather map. These lines join together places with the same air pressure, measured in **millibars** or **kiloPascals.** If the lines are close to each other, then the pressure will change a lot over a short distance. This means there are strong winds there.

If the measurement for the air pressure on the weather map is under 1,000 millibars (100 kiloPascals), this indicates low pressure. The middle of an area of low air pressure is called a **cyclone.** There will be a lot of rain and showers there. In winter, this could also mean snow is on its way.

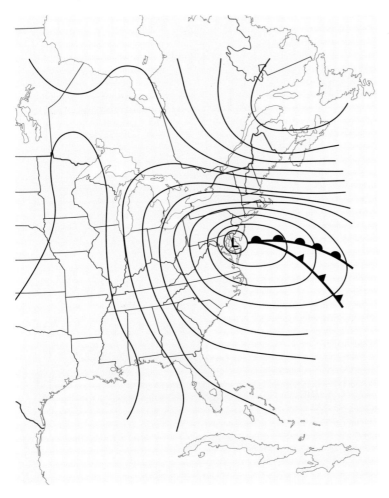

The line between two **air masses** is called a front. Triangles are used to label a cold front. Semi-circles on the lines indicate a warm front. You can see these features on this map and on weather maps on television and in newspapers.

When the measurement for the air pressure on the weather map is between 1,025 and 1,050 millibars (102.5–105 kiloPascals), this indicates high pressure. The middle of an area of high pressure is called an **anticyclone.** In the summer, when the weather map shows few isobar lines, it will be windless and settled. In the winter, the same isobars may mean fog or frost.

Fronts

Lines with triangles or semi-circles on them show where air of different temperatures meets. If there are blue triangles, it means that cold air is pushing warm air. This is called a cold **front.** A cold front forces the warm air up and brings stormy, often thundery weather. If you see red semi-circles, it means that warm air is pushing cold air. This is called a warm front. The warm air slowly rises over the cold air and brings rain.

Cold and warm air eventually mix together to make what is called an occluded front. This is where the stronger of the two fronts has overtaken the weaker front. The weather underneath will be very unpleasant and stormy.

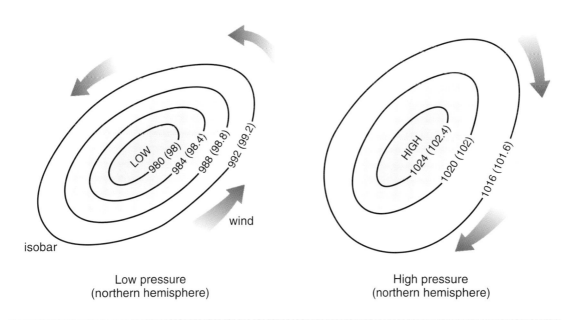

Low pressure
(northern hemisphere)

High pressure
(northern hemisphere)

In the northern hemisphere, the wind blows counterclockwise around a low-pressure area and clockwise around a high-pressure area. In the southern hemisphere, these directions are reversed. These diagrams show wind directions in the northern hemisphere. The measurements on each diagram show the air pressure in millibars (kiloPascals are in brackets).

Glossary

air mass body of air in which all the air is of approximately the same temperature and humidity

air pressure pressure at the surface of the Earth caused by the weight of the air in the atmosphere. High air pressure is a reading on the barometer of between 1,025 and 1,050 millibars (102.5 and 105 kiloPascals); low air pressure is a reading on the barometer of between 950 and 975 millibars (95 and 97.5 kiloPascals).

altitude height, usually measured from sea level

anemometer instrument for measuring wind strength

aneroid describes a barometer that does not use liquid to monitor air pressure. Instead it uses a sealed unit which has had most of the air sucked out. This is called a partial vacuum.

anti-cyclone system of winds blowing in a spiral around a center of high pressure

atmosphere gases that surround our planet

atmospheric pressure another term for air pressure

back change of wind direction in an counterclockwise direction

Beaufort Scale system of recording wind speed, devised by Francis Beaufort in 1805. It is a numerical scale ranging from 0 to 12: calm is indicated by 0, and a hurricane by 12.

climate weather conditions in a place over a long period of time

continental describes a wind that comes from a large land mass

cycle set of events that happen one after the other in the same order, over and over again

cyclone system of winds blowing in a spiral around a center of low pressure

equator imaginary horizontal line around the Earth that divides the Northern hemisphere from the Southern hemisphere

front front edge of an air mass, where it meets air of a different temperature

kiloPascal measurement of air pressure. One kiloPascal is equal to ten millibars.

knot nautical mile per hour, used in navigation and meteorology

maritime having to do with the sea

mercury silvery metal that is a liquid at ordinary temperatures

meteorologist person who studies the atmosphere in order to forecast the weather

millibar measurement of air pressure often used on barometers

monitoring station place where weather data is received and used for weather forecasts

prevailing most common or frequent

raw data information that has not been analyzed

station circle collection of symbols used by meteorologists to show local weather conditions

trend repeated pattern in the observed weather

veer change of wind direction in a clockwise direction

weather balloon large balloon filled with a gas that makes it rise through the air. These balloons carry various weather instruments that send data back to meteorologists.

weather station collection of weather instruments that measure the weather regularly

weather vane instrument used to show which direction the wind is coming from

More Books to Read

Baxter, Nicola. *Rain, Wind and Snow.* N.Y.: Raintree Steck-Vaughn, 1998.

Chambers, Catherine. *Hurricanes.* Chicago: Heinemann Library, 2000.

Chambers, Catherine. *Tornadoes.* Chicago: Heinemann Library, 2000.

Gibson, Diane. *Wind Power.* North Mankato, Minn.: Smart Apple Media, 2001.

Scholastic, Inc. Staff. *Wind and Weather: Climates, Clouds, Snow, Tornadoes and How Weather is Predicted.* N.Y.: Scholastic, Inc: 1995.

Index